The Day of the Twelve-Story Wave

The Day of the Twelve-Story Wave

Grinding Glaciers, Howling Hurricanes,
Spewing Volcanoes and
Other Awesome Forces of Nature

Diane Swanson
Color illustrations by Laura Cook

Longstreet Press
Atlanta, Georgia

Published by LONGSTREET PRESS, INC.
a subsidiary of Cox Newspapers
a subsidiary of Cox Enterprises, Inc.
2140 Newmarket Parkway, Suite 118
Marietta, Georgia 30067

Copyright © 1995 Diane Swanson

Published by arrangement with Whitecap Books Ltd.,
North Vancouver, British Columbia.

All rights reserved. No part of this book may be reproduced in any form or by any means without the prior written permission of the Publisher, excepting brief quotations used in connection with reviews written specifically for inclusion in a magazine or newspaper.

1st printing 1995

Cover and color illustrations by Laura Cook
Edited by Elaine Jones
Cover design by Designgeist
Interior design and typesetting by Margaret Ng

Library of Congress Catalog Number 95-77260

ISBN 1-56352-260-8

Printed in Hong Kong

Contents

Acknowledgments	vii
Nature Power	1
1. Heavyweight Action	3
2. The Spewing Earth	15
3. Wild, Wild Waves	27
4. Nature's Air Force	37
5. Shocking Skies	47
Index	54
About the Author/Illustrator	56

ACKNOWLEDGMENTS

I am indebted to the following for their careful and thoughtful review of various sections of this book: Rick Thomson of the Institute of Ocean Sciences, Fisheries and Oceans Canada; Jim Steele and Marian Jones of the Atmospheric Environment Service, Environment Canada; Herb Dragert of the Pacific Geoscience Centre, Geological Survey of Canada; David Skjonsberg of the Avalanche Control Section, Mount Revelstoke and Glacier National Parks, Canadian Parks Service; and Michael Roberts of the Department of Geography, Simon Fraser University, Canada.

Sincere thanks also go to Laura Cook for the remarkable illustrations she produced for this book and to my husband, Wayne, for his steady help and encouragement.

Nature Power

Nature has many powerful forces: forces that lash the sea and the sky, forces that churn within the planet and forces that press their way across the land. They build mountains, carve valleys, form lakes and more. They have greater strength and bring about more change than anything people have ever created.

Forces of nature often work together. An earthquake can trigger a snowy avalanche on mountains and giant waves in the sea. It can bring a volcano to life and, in turn, set off a spectacular display of lightning. Mighty winds can pile up ocean water. And they can whip up storms that riddle the sky with electricity and beat the ground with stones of ice.

Sometimes nature's powerful forces conflict with people. They can destroy ships, buildings, roads and crops. They can cause deaths and widespread disaster.

But there is much people can do to avoid danger and protect themselves—even against earthquakes. Many scientists think that risks to life and property would be much smaller if people were well prepared and reacted to warnings. For instance, earthquakes can suddenly damage tracks used by high-speed trains in Japan. So an earthquake warning system shuts off the electricity that runs the trains whenever gauges near the track sense extra motion. People can also try to escape danger by building houses that "ride out" the shaking of earthquakes. They can avoid beaches when strong waves race along the coast. And they can use shelters to escape powerful winds, like tornadoes.

The United Nations, an organization of countries around the world, chose the 1990s to be the International Decade of Natural Disaster Reduction. The goal was to encourage scientists and organizations from many nations to work together to reduce dangers from natural forces. Through this and other efforts, countries are getting better at predicting just where nature's forces may strike—and when. For instance, scientists are learning more about tracking pressure that causes volcanoes to explode. And that helps them give people earlier warnings.

The forces of nature are awesome, but it's wrong to think of them only as disasters. They play natural and important roles in the world, such as balancing temperatures, water supplies and nutrients. Their strength and power deserve respect. Without them, Earth would be a dead planet.

Chapter 1

Heavyweight Action

We don't usually think of solid ground, huge piles of snow or thick sheets of ice as moving objects. But under certain conditions, they all move—and forcefully. An earthquake can cause the whole planet to vibrate for weeks. Piles of snow sliding full-speed down mountains can reshape rocky peaks. And the slow grinding of massive ice sheets can bend and chisel the face of Earth.

A Whole Lot of Shaking

At incredible speeds—up to 36 000 kilometres (22 500 miles) an hour—waves of energy sometimes race as many as 600 kilometres (400 miles) up through Earth. When they strike the surface, they cause the ground to shake suddenly, rapidly.

As two of Earth's thick plates grind past each other, underground rocks snap, setting off fast-moving waves of energy.

People have always had ideas and told stories about what produces the energy to make the ground tremble. One Japanese legend tells of a giant catfish that thrashed through underground mud, shaking the surface of Earth. A story from India describes Muha-pudma, a mountain-sized elephant that held Earth on his head and caused tremors when he moved. Early Romans thought that strong underground winds caused quaking. And ancient tribes in Peru believed that a god's heavy footsteps shook the ground.

Today, scientists think that the sudden breaking of rocks is usually what sets off the energy waves that cause Earth to shake. That breaking action—plus the waves of energy and the shaking of the ground—is called an earthquake.

Many earthquakes happen along the edges of thick plates of rock that form Earth's crust. These plates glide slowly over very hot masses inside Earth, carrying the land and sea floor with them. As the plates grind past each other, they often cause rocks to break or crack.

Sometimes the plates lock together. Then pressure builds

along the edges until it is released. That can happen gradually through thousands of weak earthquakes or suddenly through a single strong one.

Other things, such as erupting volcanoes and collapsing caves, can also make the ground shake. Some large meteorites, like one that exploded 10 kilometres (6 miles) above Russia in 1908, can cause tremors. Even people sometimes cause trembling within small areas when they explode bombs or dig out large mines that collapse. And by piling up water behind huge dams, like Hoover Dam in the United States and Kariba Dam in Zambia, people increase pressure underground and cause the surface to shake.

Many earthquakes are so mild that nobody notices them. Others are so strong that land buckles and cliffs of solid rock sway. Some quakes topple mountain peaks; others raise lake beds to dry land or bury islands beneath water.

Scientists use scales of numbers to describe the movement of the ground during earthquakes. One of the best known is the Richter scale. Each number in this scale stands for a level of ground movement 10 times greater than the number below it. An earthquake of "5" is 10 times greater than one of "4" and 100 times greater than a quake of "3." People barely feel an earthquake of about "2," but a quake that has the power of 1000 nuclear bombs ranks about "8."

An earthquake in 1819 caused an island fort in India to drop three metres (almost 10 feet), forcing soldiers to escape by boat.

In 1964, during one of North America's greatest earthquakes (over 8 on the Richter scale), some shores and cliffs fell into the Pacific Ocean. Levels of land and sea floor in and around Alaska rose or

The Lost City of Kourion

Kourion was once a busy seaport in Cyprus. But in the year 365 A.D., it disappeared. Sixteen centuries passed before a young man uncovered one of the houses in the city. When researchers dug out the rest of Kourion, they learned why it had disappeared: an earthquake had destroyed it.

By sifting through clues—such as bits of tiles, wall blocks, broken pottery and skeletons—the *researchers pieced together some of the events that had happened. They think that early one morning, while the city still slept, the ground shook slightly. The tremors wakened no one, but they troubled a mule. A teenaged girl rose from her bed and went to quiet it.*

The next minute, a strong tremor alarmed the city. One family—probably like many more—woke up frightened. They huddled together as the roof of their home began to collapse. Then another, very powerful tremor struck. Walls crumbled. Buildings exploded. And the earthquake buried the whole city in its own rubble. Aftershocks continued for 50 years and Kourion became a "lost city."

fell by as much as 2.5 metres (8 feet). Like most quakes, this one caused a series of lesser shakes, called aftershocks. In fact, it produced as many as 10 000 aftershocks.

Each year, up to 22 000 quakes that rank at least 2.5 occur somewhere on Earth. That includes an average of one quake that ranks about 8. Some of these quakes have damaged towns and cities and taken many lives. And some have caused floods, fires, landslides and wild sea waves, which often do even more damage.

Besides feeling an earthquake, it is common to hear one. Many make noises like fast-moving trains. One witness to a quake in Japan described hearing the sound of a train, followed by the roar of wild animals. As the ground began to move, he heard a low groaning.

People sometimes "see" an earthquake, too. The air above the source of a quake can become charged with electricity. And that produces an eerie glowing light, called earthquake light.

As far as anyone knows, earthquakes have always occurred. Many have happened in the middle of oceans where only scientists were aware of them. But most have happened around the rim of the Pacific Ocean and in Asia and southern Europe.

One of the world's greatest quakes occurred in the Pacific Ocean off Chile in 1960. At about 8.5 on

the Richter scale, it was strong enough to shake the entire southern part of the country, rip houses from their bases and flood towns with walls of sea water. About 5000 people died.

Earthquakes hurt few people directly; falling objects, like chimneys from shaking buildings, cause most deaths and injuries. Today, many people in earthquake zones work to make cities safer. For instance, they require builders to make houses that won't easily collapse from the shaking. They also teach people to protect themselves by taking cover instantly.

A Whole Lot of Sliding

A minute is all it takes to turn a soft shuffle into a raging roar. A heavy, moving mass of snow, called an avalanche, creeps forward—just a bit—then zooms the rest of the way down a mountain slope. It often travels 80 to 300 kilometres (50 to 200 miles) an hour. One reached speeds over 450 kilometres (300 miles) an hour.

An avalanche seldom lasts more than a few minutes, but it is still one of nature's heavyweight power players. As it flows, it gathers more snow, ice—and sometimes rocks and plants—growing in weight up to a million tonnes (over a million tons). And as it grows, it gains speed. A rushing avalanche can toss boulders and trucks aside as if they were balls. Sometimes, it moves so fast it

Beating Nature to It

In the Rogers Pass of British Columbia, Canada, people start avalanches on purpose. So much snow piles up in parts of this mountain pass that avalanches barrel down 134 major pathways, threatening traffic along 40 kilometres (25 miles) of highway and railway. That's why Canada set up one of the largest avalanche control programs in the world.

As part of this program, crews built structures, like concrete tunnels, that protect the road and railway. But an avalanche control team also guards against avalanches. From mountaintop stations, the team studies weather reports and snow information. Every day, team members monitor and patrol the slopes. When the snow threatens to slide, the team stops traffic through the pass and calls in the cannons. The firing triggers avalanches and brings masses of snow down safely.

The program has paid off. Since the avalanche control team has been on the job, no snowslide has killed or injured a single traveler in the Rogers Pass.

Racing Glacier

In 1993, North America's biggest glacier, the Bering Glacier of Alaska, started to race. That's something it does every 20 years or so. But this time it reached speeds up to 100 metres (330 feet) a day—a breathtaking rate for any glacier. About half of the 5000-square-kilometre (2000-square-mile) ice mass surged forward, rumbling, cracking and buckling as it went.

"There probably isn't going to be anything like this for the rest of our careers," said Dennis Trabant of the U.S. Geologic Survey. He was one of several scientists studying the racing glacier. They gathered information from satellites and reports from bush pilots flying over the remote glacier. The scientists even set up a self-shooting camera, which took glacier pictures—until a hungry bear smashed it. They hope that studies of speedy Bering Glacier may help them learn more about just how a mass of ice can race.

flies right off the slope, creating blasts of wind that can blow buildings apart. Even when it stops, the avalanche has strength. Its snow packs down as hard as cement and holds everything firmly in place.

Avalanches take shape when heavy snowfall blankets a steep slope already buried by snow. If the new snow is very wet, avalanches are even more likely to occur. They can also form if built-up snow freezes and then melts, loosening its bond with the slope.

Masses of snow can become so unsteady that almost anything will trigger an avalanche. The shaking from earthquakes and erupting volcanoes can start them off, but so can much gentler action: the swoosh of a skier or the footsteps of a deer.

Most avalanches occur far from people. The Himalaya Mountains in Asia, for instance, probably have more avalanches than any other place. But few people ever see them.

When avalanches occur near communities or roads, however, they can cause problems. One of the worst snow accidents happened in Peru in 1970 when an earthquake triggered a huge avalanche. At speeds over 320 kilometres (200 miles) an hour, snow, ice and rock raced down a mountain and traveled up and down hills for 15 kilometres (9 miles). The avalanche flung enormous boulders, flattened villages and buried 18 000 people.

A Whole Lot of Grinding

Unlike avalanches, thick masses of ice, called glaciers, move very slowly. But they're so big and heavy, they pack enough power to grind rock. Even the smallest are about 200 metres (660 feet) long and most are much larger. One giant in Antarctica is over 400 kilometres (250 miles) long and 60 kilometres (40 miles) wide.

Glaciers shrink by melting and breaking, and grow by gathering snow and ice. When glaciers gain more than they lose, they move. Their enormous weight and the force of gravity pull them downward and push them along.

As they travel, glaciers carve out mountain peaks, valleys and inlets. They smooth and round rock—or gouge and break it. They move giant boulders and tiny pebbles long distances to new sites. They pile up hills.

Thick glaciers on steep slopes often travel the fastest. But it's hard to see even the speediest glacier move. Most don't cover 60 centimetres (2 feet) in a day. And the whole glacier doesn't travel at the same speed. Ice in the middle usually moves faster than ice on the sides and bottom where it scrapes against the ground. That's why deep cracks, called crevasses, often form along the sides of glaciers.

Sometimes water builds up beneath a glacier and creates a slippery slide. Then the ice may flow faster, suddenly increasing its speed by up to 100 times. As it moves, it creaks, groans—even booms—and its surface heaves slowly.

Some glaciers are the remains of huge fields of ice that covered much of Earth thousands of years ago. Others are much newer

In the late 1700s, visitors to the Unteraar Glacier in Switzerland were amazed. They saw a huge boulder—once carried by the glacier—still sitting on top of a mound of ice.

glaciers that formed in cold places where snow piled up and turned to ice. Even along the hot equator, glaciers grow on cold, high mountaintops. But a glacier takes a long time, often hundreds of years, to develop.

* * *

Shaking ground. Sliding snow. Grinding ice. People benefit from these heavyweight forces. The carving done by glaciers and avalanches, nature's sculptors, makes Earth more beautiful. And glaciers store about three-quarters of the world's fresh water. Some supply water and generate electricity for cities and farms in countries such as the United States. Even earthquakes are useful. They help us learn more about how mountains are formed, how continents move and how the inside of the planet works.

Opposite: **Glaciers sometimes flow to the sea where chunks may break off as icebergs.**

Chapter 2

The Spewing Earth

We see only the surface of our planet so it's easy to forget there is an "inner Earth." But sometimes it reminds us. It bursts through openings and cracks in the surface, spewing out masses of material. Volcanoes can release enough melted rock to bury hundreds of farms. Fountainlike geysers shoot out water and steam with such force they can make the ground tremble. And some slits in the floor of the sea eject fountains of water so hot they can melt thermometers.

Marvelous Mushrooming Mountains

Deep inside Earth flows a fiery mass of hot and melted rock, called magma. Where this magma wells up and breaks through weak spots at the planet's surface, volcanoes rise. They are mountains built by

Hot and melted rock breaks through a weak spot in Earth's surface, building a mountain.

the outpouring of magma—or lava, as it's called when it reaches the surface.

No two volcanoes are alike. Their openings vary from vents just a few metres (several feet) wide to craters several kilometres (several miles) across. Some volcanoes are cone-shaped; some are flat-topped. And they can change shape—fast. One volcano in Russia formed a cone 600 metres (2000 feet) tall in only four months. But volcanoes are often unstable and their cones can collapse in landslides.

Volcanoes may be active—erupting masses of ash, gas or lava—or they may be inactive and then start erupting again years later. Most erupt off and on for about a million years before they finally die.

Sometimes eruptions from volcanoes blow masses of ash high enough to clog the engines of airplanes. And eruptions can form a long-lasting haze that blocks out some of the sun's heat. After one of the world's largest eruptions occurred at Mount Tambora in Indonesia in 1815, thick haze caused a "year without summer." Heavy snow fell on Europe and North America in June and harsh frosts attacked crops in July and August.

Like earthquakes, erupting volcanoes can trigger floods by damming rivers and causing lakes to overflow. Volcanoes that erupt beneath thick masses of ice, or glaciers, can also cause floods. If all the ice above the volcano melts, a lake forms. Sooner or later, this lake breaks through the volcano's rim, drenching the land and dotting it with icebergs. In 1918, Volcano Katla in Iceland erupted for two days. It released more lake water every second than the water carried by the Amazon, the world's largest river.

Eruptions can hurt or kill people who live near volcanoes. That's why researchers are working to improve forecasts of eruptions. By tracking the movement of lava and checking the pressure buildup from gas, they can make better predictions. Earlier warnings make it easier to move people to safety.

But most volcanoes rise far away from people. Some form in space—on moons and planets. Astronomers have seen volcanoes erupting on Io, a moon of planet Jupiter. The Olympus Mons on planet Mars is thought to be the largest volcano in the solar system. It stands 27 kilometres (17 miles) tall and 700 kilometres (450 miles) wide.

On Earth, most volcanoes form in the sea, usually in clusters or chains. In the Pacific Ocean alone, there may be as many as 14 000. Some volcanoes poke above the surface, forming islands, like those of Hawaii and the Philippines.

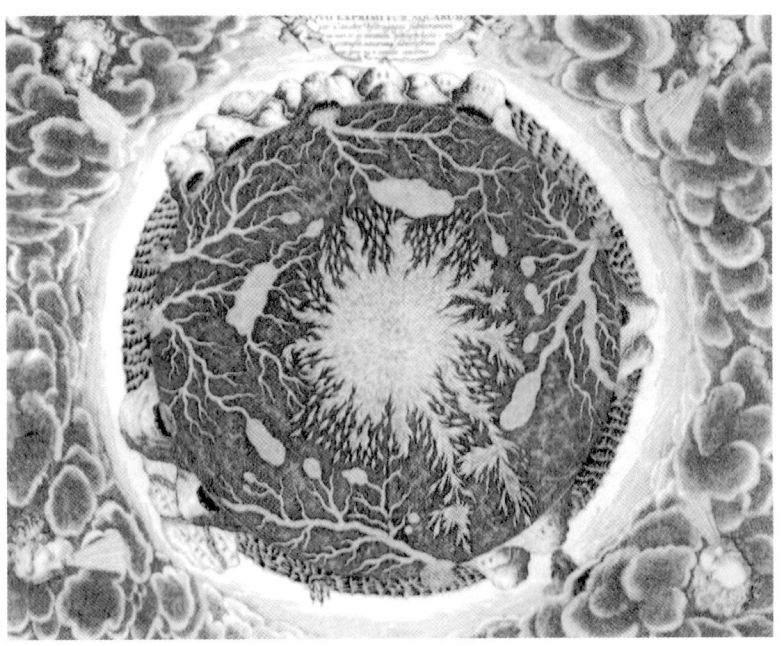

A 17th century view of the inner Earth shows fire feeding volcanoes and heating geyser water.

17

The Power of Pinatubo

In 1991, Mount Pinatubo in the Philippines blew, causing one of the biggest volcanic explosions of the century. It spewed clouds of hot gas, masses of lava and a tower of ash 40 kilometres (25 miles) tall. Daylight became darkness. Lightning pierced the sky with flashes of blue, green and red. And the outpouring of lava created a large underground hollow that collapsed and caused the ground to shake. Losses included 42 000 houses, 40 000 hectares (100 000 acres) of crops and 900 lives.

Although Pinatubo hadn't exploded for 600 years, steam from a few vents spurred scientists to measure movement of magma under the mountain. Four days before the eruption, they knew the mountain was ready to blow, and crews moved 200 000 people out of danger.

But the effects of Pinatubo's blast went beyond the Philippines. Astronauts in space said they had never seen Earth look so hazy. Within three weeks, a thin cloud had encircled the planet, reducing sunlight and temperatures. During the two years that followed, average temperatures, worldwide, fell below normal.

In the Atlantic Ocean, the island country of Iceland was also built by underwater volcanoes. In 1963, Icelanders witnessed the birth of another island. They watched the sea boil from the eruption of an undersea volcano. By the next night, a ridge of land rose above the surface. About a year later, that ridge had become an island 1.5 kilometres (1 mile) wide. Icelanders named it after Surtsey, the god of fire.

But most undersea volcanoes remain completely covered by water. Even the peak of Great Meteor Seamount, the 4200-metre-tall (14 000-foot-tall) volcano off Africa, is about 270 metres (900 feet) below the surface. People usually don't even know when an underwater eruption occurs, but sometimes they witness the ocean's response to one.

Sailors on a cargo ship off Indonesia in 1993, for example, felt three shocks when an underwater vent erupted. Then they saw a plume of steam rise above the sea's surface. Two explosions followed and each time a tower of water 100 metres (330 feet) high soared into the air.

That same year, scientists were able to watch the eruption of an undersea volcano—the first eruption anyone had ever seen in the ocean. A submarine tracking station had recorded earthquake noise along the west coast of the United States, alerting Canadian and American researchers. They arrived within days of the start of the eruption two kilometres (over 1 mile) beneath the sea.

Sending down a remotely controlled submarine with a video camera and other equipment, the researchers focused on the sea floor. The camera showed a newly opened split—about 7 kilometres (4.5 miles) long and 100 metres (330 feet) wide. Lava was spewing out and a snowstorm of bacteria was settling in thick mats all over it. Scientists aren't sure if the bacteria rose from deep beneath the vent or if they lived around it. But the bacteria thrived on the chemicals and minerals that came from the vent.

Great Gushing Geysers

All is quiet. Then there's a rumbling. A burbling. A sudden, thundering explosion. Shooting straight up, hot water gushes from a tall cone and steam billows out for an hour or more. Slowly, the air clears. All is quiet again.

Geysers are one of Earth's showiest forces. Spewing water and steam high into the air, they are nature's own fountains, and many of them are beautiful. Some are major attractions in parks.

It's easy to understand how a geyser works. Just think about the water heater in a house. It has a source of heat, like gas or electricity. It has a tank to store and warm up the water. And it has openings to let hot water out and cold water in.

A geyser has these things, too. Its source of heat is hot rocks that are deep underground. It has a hollow, or reservoir, beneath the surface to store and heat water. It has an opening to eject hot water and underground channels to bring in water that has seeped into the ground.

When the Great Geysir in Iceland was active, it used to shoot water nearly 30 metres (100 feet) high.

As the water in the reservoir heats, it creates steam near the opening. Steam needs more space than water so it ejects the water. Some steam also escapes. Then there's room to take in more water from underground channels, and the cycle starts again.

Hardly anyone sees inside a geyser. But in New Zealand, a man climbed inside one that was no longer active. Its opening was just big enough for him to squeeze through. Dropping down about five metres (16 feet), he found himself in the geyser's reservoir. It was about three metres (10 feet) high and four metres (13 feet) long. One of the channels in the floor of the reservoir connected with an active geyser. The man could hear it rumbling with boiling water.

Although no two geysers are exactly the same, there are two main types: cone and fountain. Cone geysers have cone-shaped openings that create tall, narrow sprays of water. Fountain geysers have bowl-shaped crater openings that produce broader, fountain-like sprays. Water builds these cones and craters as it cools at the surface, depositing minerals in layers.

Like volcanoes, some geysers are active and some are dead. Others may be active, inactive, then active again. Some always gush water about the same time and the same way, but many do not. One famous American geyser, called Old Faithful, is almost as regular as its name suggests. It usually shoots a 50-metre (160-foot) spray of water and steam every 50 to 80 minutes.

Geysers occur around the world in areas where volcanoes exist. But these geysers only number about 400. Most of them are in the United States, Iceland, New Zealand, Russia and Japan. The largest on Earth is Waimangu Geyser in New Zealand. Although it hasn't been active since 1917, it used to erupt every 36 hours or so. Water in its 100-metre-wide (330-foot) opening would boil vigorously and

Opposite: **Masses of hot water and steam spurt out of the Earth, forming geysers of great beauty.**

Seeing Is Believing

In the mid-1800s, guides and trappers from the western United States told stories of an awesome sight: a field of about 200 steamy, water-spouting geysers. But many people in the eastern states thought these stories were just make-believe.

In 1871, a photographer piled 90 kilograms (200 pounds) of equipment onto a mule and joined researchers heading west to the Yellowstone area of Wyoming. There he photographed the largest field of geysers on Earth. His pictures helped the researchers convince the government that geysers really existed. The next year, the United States created its first national park to protect these geysers: Yellowstone National Park.

then the geyser would suddenly spew water, steam, mud and rocks 450 metres (1500 feet) into the air. Minutes later, all would be quiet once again.

Stunning Steamy Sea Floors

Deep in the ocean is another kind of water-gushing geyser: a hydrothermal vent (*hydro* means water and *thermal* means heat). Sea water seeps down through cracks in the ocean floor where hot rocks heat it. Then it expands and rises, flowing out as warm streams or blasting out as scalding water at speeds up to two metres (7 feet) a second.

Some hydrothermal vents are low-lying. Others have chimneys, built from minerals that settle out when hot water cools. Called black smokers because they spew clouds of dark, mineral-rich water, these chimneys often grow to 20 metres (65 feet). Some are as tall as 50 metres (160 feet).

Doing research at Canada's University of Victoria, Kerry Wilson described his first visit to Pacific Ocean vents: "We descended for an hour in the submarine to depths of about 1.6 kilometres (1 mile). Then we began to fly across the ocean bottom. It was like a desert—dark and barren. Then, suddenly it was right in our lights: a chimneylike vent standing about five metres (16 feet) tall. Just beyond was a whole field of vents. It was like something from a dream."

Although it's likely that hydrothermal vents have always been on ocean floors, no one saw any before 1977. That was the year researchers first discovered vents in the Pacific Ocean—under 2500 metres (1.5 miles) of water, far off the coast of South America. Scientists had been trying to learn why the temperatures in that sea water were much higher than in other areas. They searched the sea floor,

Worms, crabs, mussels, and other sea life thrive in the darkness at deep-ocean vents.

first with cameras, then with a submarine. And they made their stunning discovery: a fantastic field of warm-water fountains.

Later, in 1985, researchers found vents in the Atlantic Ocean—in water five kilometres (3 miles) deep. After crossing empty stretches of dark sea floor, they found a field splattered with color: red, yellow and green chunks of minerals. Black smokers were jetting hot water that was rich with minerals. Scientists even found gold—something they once thought only built up on land.

* * *

The spewing Earth helps us in many ways. Volcanoes raise mineral-rich magma from the inner planet and make it available in soil to help plants grow. They build up land, both above and below the sea. In fact, millions of years of volcanic eruptions formed the sea floor.

Geysers provide hot water to heat buildings and steam to power machines that create electricity. The heat and minerals in geyser water also help people relax and improve their health.

Studies of black smokers undersea help scientists learn how minerals form. And the plants and animals that live there may provide clues about how life survives in darkness, how it existed ages ago—and even how it might exist on other planets.

There's Life Down There

Giant clams, huge mussels and ghostly white crabs. Masses of tube worms up to 1.5 metres (5 feet) long. Scientists who found hot-water vents in the floor of the Pacific were amazed at the life there. They could hardly believe these animals existed in the darkness of the deep sea.

But water from the vents is rich in hydrogen sulfide, a chemical that some bacteria seem to use to make food. In turn, these fast-growing bacteria provide a lot of food for other life in the food chain. In fact, the food supply is 500 times greater at some vent fields than it is at deep-water sites without vents.

In the Atlantic Ocean, five-centimetre-long (2-inch) shrimp swarm on chimneylike vents to eat bacteria. Scientists discovered sensors on the backs of these shrimp that seem to pick up dim light from the vents. No one knows what causes this light, but researchers wonder if bacteria can grow by using it in place of sunlight. And the mysteries of deep-sea vents continue.

Chapter 3

Wild, Wild Waves

In the Atlantic Ocean off Scotland, waves swept away a pier that weighed as much as 22 fully loaded train cars. In the Pacific Ocean off Japan, a single wave flung a 770-tonne (850-ton) block of coral two kilometres (over 1 mile) through the air. Although some waves just ruffle the sea, others grow so big, so fast and so strong they rank among Earth's most powerful forces.

Whatever their strength, waves are simply moving ridges of water. In the ocean, they form whenever the sea is disturbed. Winds blowing across the water cause most waves. But major uproars, like underwater earthquakes and eruptions from volcanoes, can cause very strong waves, called tsunamis (say "soo-nom-eez").

When the Wind Blows

Moving across oceans, winds pile up water into waves. Generally, the stronger the wind, the greater the number of waves and the taller they are. Light puffs of wind, called "cat's paws," only ripple small patches of the sea's surface. But fast-moving winds that blow for days across hundreds of kilometres (hundreds of miles) of ocean whip water into wild waves.

Most windblown waves at sea are less than 4 metres (13 feet) tall. An average of 1 in about 300 000 rises 15 metres (50 feet)—as tall as some 5-story buildings. Still, even higher waves have been recorded. These giants, called rogue waves, form when several waves pile up instead of following each other one by one. They can form when groups of large waves try to run against strong ocean currents, like those off Japan and the southeast coast of Africa. Rogue waves usually form without warning, but last just a minute or two.

The tallest rogue wave officially recorded was 34 metres (112 feet) high. In February 1933, a navy tanker, named the USS *Ramapo*, rode this giant during a hurricane in the North Pacific.

Wind-driven waves lash a British seawall, spewing thick, white foam.

When the Earth Shudders

Undersea volcanic eruptions, earthquakes, landslides—even meteorites—can move masses of water, sometimes producing tsunamis. These waves head out in all directions from whatever creates them. They're like water that moves in rings from rocks tossed into ponds. Although some people call them tidal waves, tides have nothing to do with causing them.

Earthquakes in the sea can cause tsunamis when the quakes rank at least 7 on the Richter scale (see page 5). They must also cause the sea floor to slip up and down—not sideways, which is more usual. Up-and-down movement of the floor disturbs the sea surface—just as volcanic eruptions and landslides do. Movements from earthquakes, volcanic eruptions and landslides often combine to cause tsunamis.

Crashing into oceans, meteorites from space can also cause tsunamis. Scientists think a meteorite struck the Atlantic Ocean 66 million years ago and created 50-metre (160-foot) waves. They raced across a sea that covered what is now part of Texas in the United States. Rocks on that former sea floor look like rocks that have been scoured and broken by tsunamis.

In deep water, tsunamis are often hundreds of kilometres (hundreds of miles) long but they seldom rise higher than a metre (about 4 feet) above the surface. People in planes and boats pass right over tsunamis without noticing them. But as these waves reach shallow water near land, they slow down and often rise much higher. In 1946, sailors on very calm water just off Hawaii were shocked to see huge waves suddenly rise up at shore and burst over warehouses.

Tsunamis normally travel fast. One can sweep from Alaska to Antarctica in less than a day. During a 1964 tsunami, which followed one of North America's greatest earthquakes, waves traveled down the coast from Alaska at an average of 530 kilometres (330 miles) an

A tsunami caused by an earthquake in the Caribbean Sea in 1867 struck a mail ship called *La Plata*.

A strong earthquake that moves the sea floor up and down produces racing waves that rise higher close to shore.

hour. But as they moved through deeper water across the open ocean to Hawaii, they averaged 830 kilometres (520 miles) an hour—about the speed of a passenger plane.

Tsunami waves pack a lot of energy—even when traveling a long way. They lose some energy by rubbing against the sea bottom and bumping into things, like islands and reefs. But these waves lose their energy very slowly.

When tsunamis reach shore, each behaves differently. One might flow in gently; another might flood the land—especially if it arrives at high tide. One might cause the water near shore to withdraw, making a loud sucking sound and exposing more of the sea bed than ever before. Another might rise in huge, heavy walls of water, then come crashing down. Many tsunamis combine behaviors, producing a gentle wave or two, then one so strong it pushes ships down city streets.

The type of shore affects the way tsunamis behave. Bays channel waves, usually making them taller. Barriers, like sandbars, often make waves shorter. In fact, the same wave may be more than 15 metres (50 feet) tall at one point but only 1.5 metres (5 feet) at another point along the shore.

Two-thirds of the world's tsunamis happen in the Pacific Ocean, especially around its rim—where earthquakes and volcanic eruptions are common.

Wave Supreme

What is called "the greatest splash in history" was neither a wind-blown wave nor a true tsunami. It was an odd event that occurred July 9, 1958—a day that seven-year-old Sonny Ulrich had spent fishing with his dad. That evening, the Ulrichs steered their 12-metre (40-foot) boat, the Edrie, into Lituya Bay, Alaska, and went to bed. But at 10:17 P.M., they awoke to an earthquake tremor so strong it shook the mountains around them. Two-and-a-half minutes later, they saw millions of tonnes (millions of tons) of mountain and glacier break off and plunge 800 metres (2600 feet) to the narrow bay.

At once rose an enormous wave—a wall of water that stretched across the bay. Traveling 160 kilometres (100 miles) an hour, the wave raced to shore, scouring cracks in rocks and scraping trees off slopes as high as 520 metres (1700 feet). Up, up, it carried the Edrie until the boat slipped over the back side of the wave and headed for open sea. Sonny and his dad survived.

Gohei: The Hero

On December 24, 1854, an old man named Gohei felt the ground tremble. Looking down from his home on a hill in Japan, he noticed the sea pulling away —far away from shore. "A tsunami!" he thought fearfully, and he knew he must act fast. About 400 people lived below him in the village of Hiro, Wakayama. Their lives were in danger.

Grabbing a torch, Gohei raced into his field and set fire to several sheaves of rice. The villagers spotted the flames at once. They ran up the hill to help put out the fire. But when they got there, they saw the sea rise very high and come crashing down. They heard a noise like thunder—but louder than any thunder they had heard before. Again and again, waves lashed the little village, then swept it out to sea.

The people watched breathlessly. They realized Gohei had saved their lives.

Each year, an average of one tsunami causes minor damage somewhere in the Pacific Rim; one in ten years causes major damage.

Yet it was the Indian Ocean, not the Pacific, that produced one of the wildest tsunamis on record. On August 27, 1883, an undersea volcano, called Krakatoa, erupted, causing one of the biggest explosions ever made. Although the volcano was in Indonesia, people in Australia—more than 4500 kilometres (3000 miles) away—heard the blasts.

From its peak, which formed a small island, Krakatoa spewed lava, rock and steam into the ocean. The force set off a mighty tsunami. Waves more than 12 stories high sank thousands of boats and planted one in a forest three kilometres (2 miles) from shore. The tsunami crashed over islands, washing many small ones away. It destroyed more than 300 communities and killed about 36 000 people.

Traveling 640 kilometres (400 miles) an hour, the tsunami then sped around the southern tip of Africa into the Atlantic Ocean and raced on. Scientists think it circled the globe two or three times before it ran out of energy.

Some of the most powerful tsunamis in the twentieth century started their travels near Chile in May, 1960. Earthquakes and underwater landslides created waves that destroyed towns on long stretches of coast. And the waves traveled

33

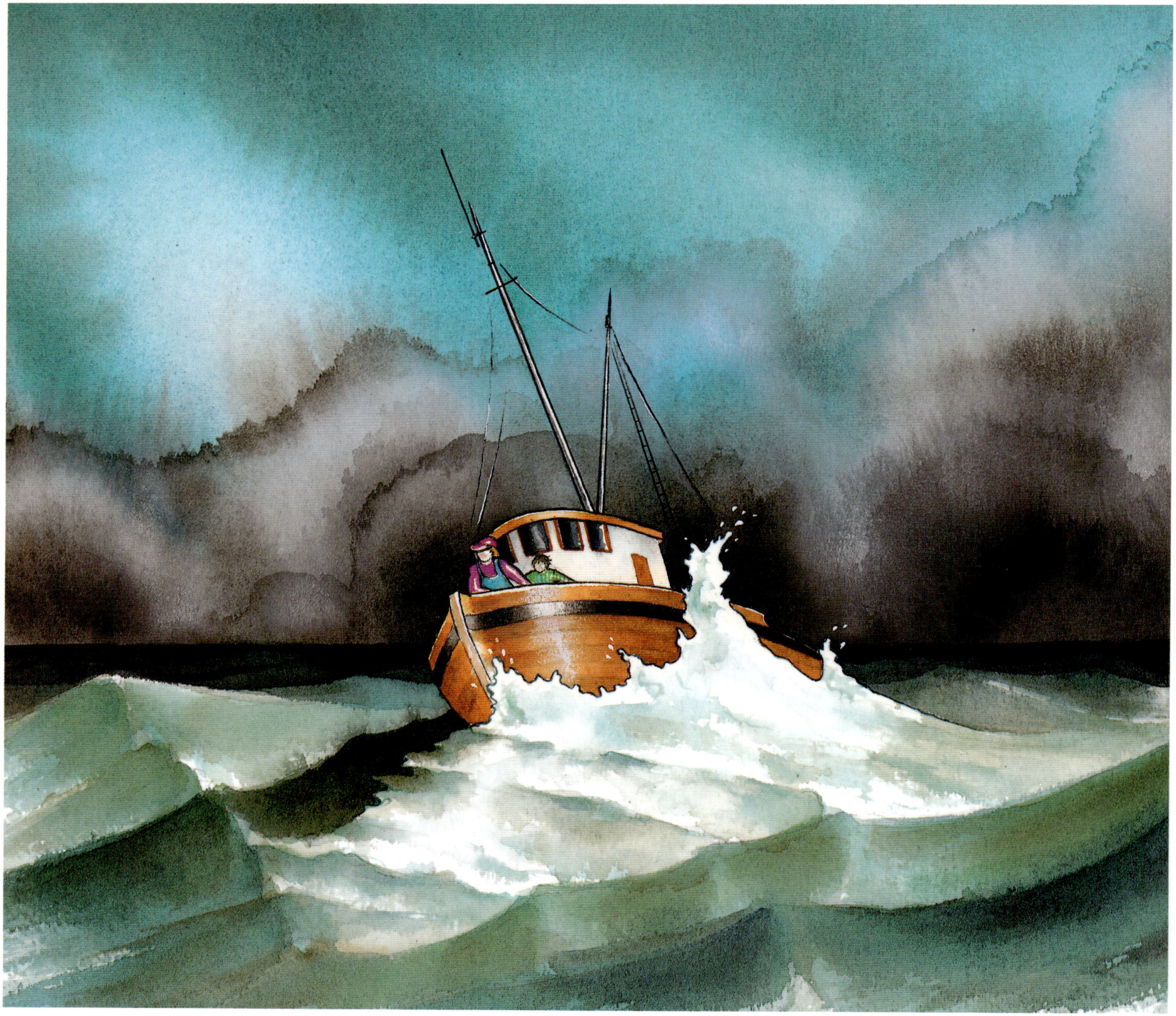

on. As they hit Hawaii, the Philippines and Japan, they took even more lives.

At Hawaii's Hilo Bay, people reported seeing and hearing waves pound the shore, grind buildings together and snap off power poles. The tsunami smeared city streets with mud and fish; it stacked up cars and wound some around trees. And it tossed 20-tonne (22-ton) rocks from a seawall into a grassy park 180 metres (600 feet) from shore.

* * *

The wild, wild waves of the ocean deserve our respect. By being aware of their power, we can try to avoid disaster. We can even try to benefit. Scientists are studying the use of wave energy to produce electricity. And surfers—well—just ask how they feel about windblown waves.

Opposite: **Strong winds build tall, fast-flowing waves, rocking boats at sea.**

Warnings That Save Lives

"*This is a tsunami warning. You are advised to leave,*" *blared the van's loudspeaker. Firefighters and police officers raced from door to door, hollering: "Head for high ground!" And people in low-lying coastal towns snatched a few valuables and fled.*

A major earthquake had just occurred. And the international Tsunami Warning System in the Pacific had leaped into action. Using the latest data on the earthquake and on changes in sea levels, it had figured out where a tsunami might strike—and when. Then it had warned the countries along those coasts so local emergency teams could warn the people.

Twenty-six countries around the Pacific Ocean take part in the Tsunami Warning System. They operate stations and gauges that record unusual changes in sea level and major disruptions, like earthquakes. The Tsunami Warning System receives this information around the clock. Since it began in 1948, it has saved many lives.

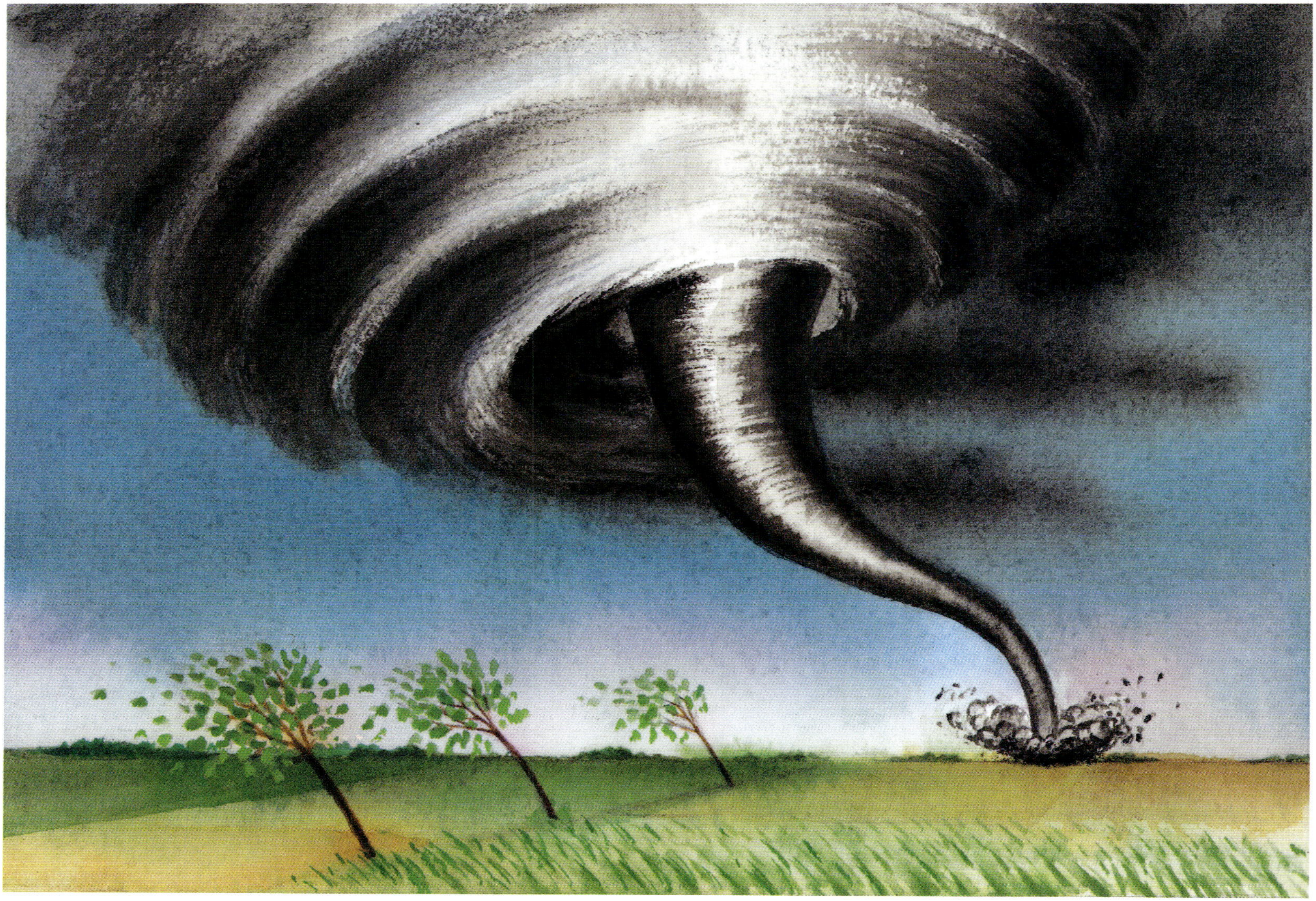

Chapter 4

Nature's Air Force

Winds are a simple force—just moving air. They are born when cool air moves in to replace warm air that rises. Over short distances, big differences in air temperatures mean stronger, faster winds.

But as simple as they are, winds are a powerful force on Earth. Some swirl in storms called hurricanes that release as much energy each minute as hydrogen bombs. Others spin at top speeds in twisting storms that can raise whole buildings, turn them around and set them down again. Other kinds of winds swoop over mountains, carrying enough heat to turn snow-buried fields into lakes within hours.

Ancient Greeks believed that Aeolus was a god of wind who stored his powerful forces in a cave.

Hulking Howlers

The biggest, most powerful storms on Earth are hurricanes. Some countries call them typhoons or cyclones, but they are all tropical storms with winds that blow at least 120 kilometres (75 miles) an hour. They last about nine days and travel up to 5000 kilometres (3000 miles). In their paths, they pile up sea water, toss ships and bulldoze coasts. If they cross land, they can destroy buildings and roads and cause floods.

Worldwide, hurricanes are born in seas near the equator—usually in summer and fall. That's when water surface temperatures are often at least 27°C (80°F). The air over the seas becomes hot and moist. It rises, cools and forms clouds. Then as more air moves in to take its place, winds arise. They whirl and howl, drawing up more hot, moist air. The clouds gather masses of water, then they dump it in dense walls of rain. One 1928 hurricane in Puerto Rico dropped 2500 million tonnes (2750 million tons)—only a fraction of its water—in just two hours.

A hurricane is huge. It can grow to be 800 kilometres (500 miles) across. Its heavy winds and rains spiral around a giant tower of calm air, called the eye, which is usually about 20 kilometres (12 miles) wide. The eye puckers the sea, forming mounds of water several metres (many feet) high. And it sucks in plants, insects, birds—even small frogs. Trapped in this fast-moving "cage," plants and animals frequently travel long distances before the hurricane weakens and lets them go. Some people suggest that hurricanes might be the cause of stories about fish raining down on farms and cities in many countries.

When sailors meet a hurricane at sea they first notice the wind rising—often after a day of calm. Then a thick, black curtain of cloud moves in and encloses the ship. Heavy rain beats down and the wind gathers strength fast. Huge waves break over the ship as the storm tosses it for hours.

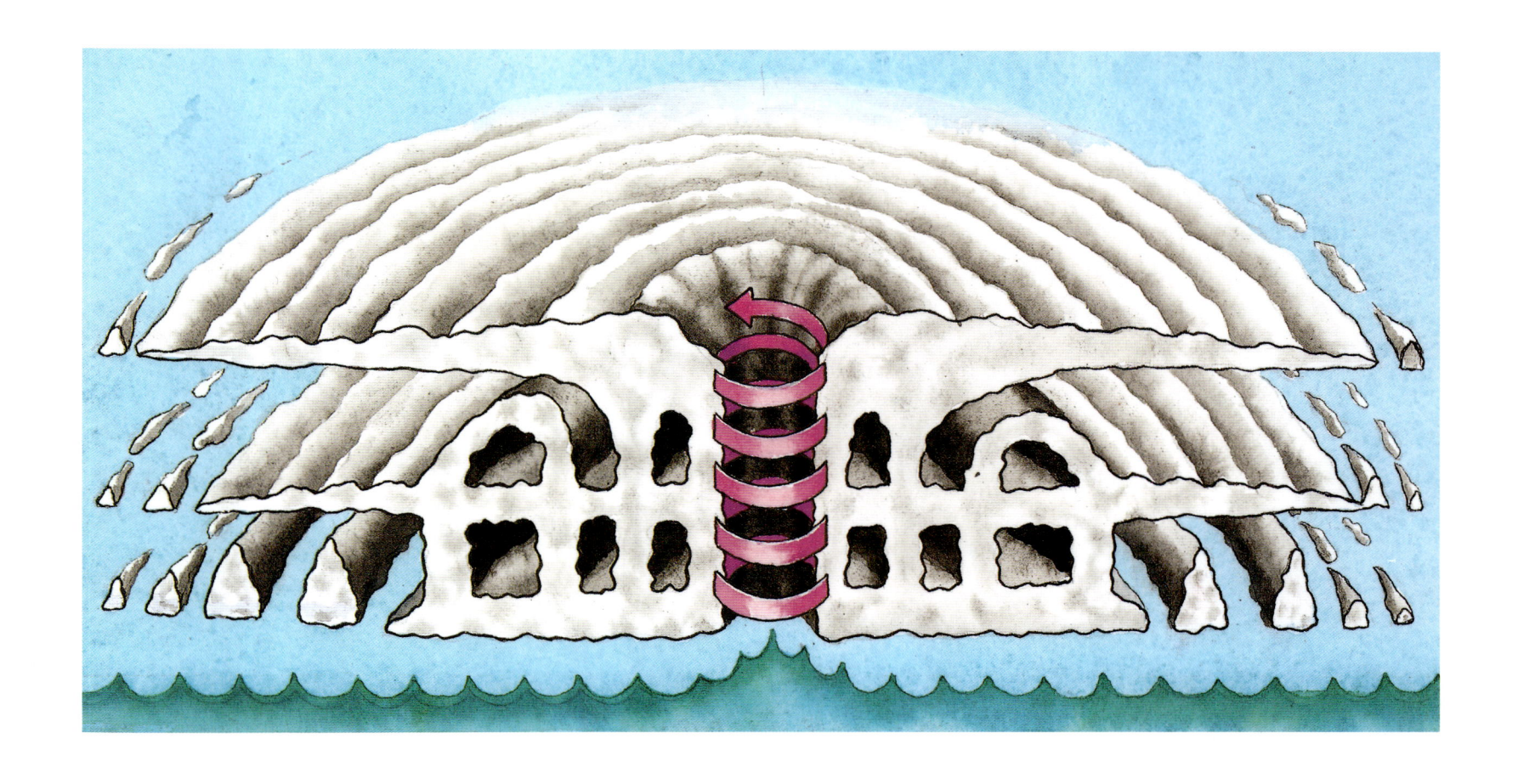

Suddenly, the sailors enter the calm eye of the hurricane and they look up to blue sky. The winds die but the sea rises in hills of water that threaten the ship. Thirty minutes later, another black curtain of cloud appears and high-speed winds return. After several hours, the air gradually grows still, clouds scatter and the ship sails on in fair weather.

A huge tropical storm forms banks of heavy rainclouds over the ocean. Powerful winds spiral around a tall column of calm air that draws sea water upward.

Hurricane Hunters

Shaking and tilting, a small plane fights its way through a swirling wall of fast-blowing wind. Then it bursts through a wall of clouds 10 kilometres (6 miles) tall and strikes dead calm. Above, sunlight shoots down from a clear sky. Below, waves loom large, then crash back into a foaming sea.

Flying into the eye of a hurricane is one way to learn about it. A skilled pilot, called a hurricane hunter, fights heavy winds to enter the hurricane's calm eye. On each trip, the pilot gets more information about the size of the storm, the speed and pattern of the winds, the temperature, the hurricane's position—and more.

Besides helping scientists understand hurricanes better, hurricane hunters help forecasters. These forecasters use data from the hunters as well as satellites to track the form and movement of hurricanes. Forecasters are usually better at judging where the storm might strike than how strong it will be. But as they learn more about hurricanes, they will be able to give clearer and earlier warnings. That means saving more lives.

Using planes, satellites, radar and computers, people who track hurricanes rate the strength of these storms. They use a scale of numbers, called the Saffir-Simpson scale. It ranges from 1, which is just a weak storm, to 5, which is the strongest storm. About 65 hurricanes occur on Earth each year. Normally, only one that ranks 5 appears in 100 years.

A hurricane needs a source of hot, moist air to stay alive. If it curves away from the equator or if it strikes land, it loses that source. Heavy rainfall releases some heat to keep the hurricane going for a while. But as it travels on, the air that moves in is much cooler. Soon, the hurricane dies.

Towering Twisters

Fast-spinning cones of air, called tornadoes or twisters, are mighty storms that contain Earth's fastest winds. Compared to hurricanes, they are narrow (up to 360 metres or 1200 feet across) and live short lives—often less than an hour. But they have the power to yank out trees, suck streams dry and rip roofs off buildings.

Tornadoes form quickly when masses of cool, dry air clash with masses of hot, moist air. Looking like whirling cones or towers, tornadoes develop from thunderstorm clouds, growing until they touch ground. They spin so fast they look almost solid but, like hurricanes, they have calm cores, or eyes.

They act like giant vacuum cleaner hoses, sucking up dirt. That's what colors many of them black.

Roaring like thunder, tornadoes can be frightening. If they cross cities and towns, they can do a lot of damage. But most tornadoes cross countrysides where few people live. They skip and jump through fields, leaving round marks as "footprints" wherever they touch ground. And they often do strange—sometimes funny—things: shear wool off sheep, pluck feathers from chickens and drive grass into tree trunks. They suck up whole ponds—frogs, lily pads and all—and dump them into farmyards. They airlift cattle from one field to another and blow cartons of eggs hundreds of metres (several hundred feet) without cracking them.

Tornadoes that form over the sea sometimes suck up fish. Called waterspouts, these tornadoes look white from the spray of the water they vacuum up.

Tornadoes form in many countries, but they are especially common in the United States. In fact, a band of flat land, called Tornado Alley, in central United States gets several hundred a year—many more than any other place on Earth. That's because moist, hot air from the Gulf of Mexico often crashes into cool, dry air from the west.

Tornado watchers know that twisters usually happen on hot afternoons in spring or summer. These storms frequently appear from the southwest, and rain or hail often falls ahead of them. Yet it is very hard to predict when one will strike—or where. Most tornadoes spend just a few seconds in one spot. They keep moving until they

Strong winds can stop a horse in its tracks.

Hot chinook winds melt snow, exposing winter food for pronghorns and other wildlife.

finally run out of heat and moisture. Until then, weather bulletins in the area warn people to take shelter.

Super Swoopers

Some of the strangest winds on Earth bake apples on trees and skim ice off ponds. These hot, dry winds, called foehns (say "phones" or "phons"), blow through several countries—in places where air rises to cross tall mountains.

When the air climbs, it cools and drops moisture as rain or snow. Then it crosses mountain peaks and swoops down slopes to valleys or plains. As it drops, the wind becomes warmer. Extra sunshine on the downslope side of mountains also raises the air temperature a bit.

Foehns arrive year-round. In spring and summer, they blow away soil and they dry plants, sometimes turning leaves to dust. In fall and winter, they "eat" snow, turning drifts and ice into vapor. Sometimes, foehns cause floods by suddenly melting masses of snow and ice.

Many foehns start with a few cool gusts, followed by a sudden calm. Then hot, dry winds blow furiously. Sometimes, they last two or three days, occasionally reaching speeds of 160 kilometres (100 miles) an hour. They can break trees, lift roofs off buildings or knock people over.

In North America, foehns, called chinooks,

Eye to Eye with a Tornado

On a hot afternoon in June 1928, American farmer Will Keller of Kansas spotted a tornado speeding his way. He raced to his underground shelter. But just before he slammed the door shut, he turned and looked back. The tornado had lifted off the ground.

Will watched as it hovered directly overhead. He smelled something strong, like gas. He heard screaming and hissing: tiny tornadoes were forming and breaking off the bottom of the large one. Then he stared up through a wide opening—right into the tornado itself. It looked like a round tower with clouds for walls, and it stretched almost a kilometre (half a mile) to the sky. Inside, lightning flashed again and again.

The tower moved on. And Will Keller became one of very few people who have looked into a tornado and lived to tell about it.

Now That's a Wind!

People in southern Alberta, Canada, enjoy telling tall tales about chinooks. One favorite is the story of a man who hitched his horse to a sleigh and headed for the next town. On the way, he heard a howling behind him. A chinook was blowing his way. Faster and faster ran the horse. Harder and harder blew the wind, melting the snow that covered the dirt road.

"Try as I might, I could just keep the front of the sleigh on the snow," said the man. "The back was scraping along the bare road, raising a thick cloud of dust."

Another story tells of snowdrifts so deep they buried a church. People had to tie their horses to the steeple and dig through the drifts to attend service. But while they were in church, a chinook whipped the snow away. When the people stepped outside, they found their horses swinging from the steeple.

zoom down mountain slopes after moist winds from the Pacific Ocean drop their moisture. On January 22, 1943, they raised temperatures in South Dakota, United States, from –20°C to 7°C (–4°F to 45°F) in only two minutes.

But chinooks blow most often in southern Alberta, Canada—on as many as 35 days each winter. In the western sky, an arch of clouds, called the Chinook Arch, signals the coming of a chinook. Its hot blasts can save starving wildlife by uncovering grass buried beneath the snow. One chinook managed to melt more than 100 centimetres (40 inches) of snow in eight hours.

* * *

Strong winds have the power to cause a lot of damage, but they have important roles to play. Hurricanes, for instance, help balance temperatures by moving heat away from the tropics. They also bring water to dry lands. Along with tornadoes, they distribute masses of seeds and plants. And foehns help ripen fruits in the fall and make it possible for many animals to survive the winter.

Chapter 5

Shocking Skies

About 2000 thunderstorms are booming over Earth this minute. They are searing the skies with lightning and filling the air with thunder. They are beating the ground with rain or hammering it with hail.

Thunderstorms are a mighty force. The giant strokes of lightning they produce are more than five times as hot as the surface of the sun. And the icy stones of hail that thunderstorms launch can strike so hard they bury themselves in the ground.

In a Flash

Lightning is a giant spark of electricity that travels 140 000 kilometres (90,000 miles) in a second. That's more than 50 million kilometres

(30 million miles) an hour. If lightning could travel around the planet, it could make the trip in less than half a second.

Bolts of lightning are so big people can see them 300 kilometres (200 miles) away. The length of the bolts usually ranges between 50 metres (160 feet) and 30 kilometres (20 miles). But astronauts in space have reported seeing some that stretched 160 kilometres (100 miles) across the sky.

Scientists aren't sure what brings about the build-up of electricity that causes lightning. But events such as volcanic eruptions create lightning when electricity mounts up in the towers of ash they spew. Most lightning occurs, however, when thunderclouds build up the difference in electrical charges between the top and bottom of a cloud—or between the cloud and the ground.

Lightning usually travels within a cloud—from one part to another. Called sheet lightning, it brightens the sky with broad flashes of light. But sometimes a giant bolt leaps between two different clouds, or between a cloud and the ground. Then it is called fork lightning because it "forks" off into two or more paths. It appears to zigzag in sharp strokes, but it really travels in smooth curves.

As bolts zoom across the sky, they heat the surrounding air, which expands suddenly. That causes the sky to boom with thunder. Light travels much faster than sound, so people see lightning before they hear thunder. They count the number of seconds between the two and divide by three to find out how many kilometres away the lightning is. (They divide by five to find out how many miles away it is.)

Lightning has cooked potatoes still in the ground.

Lightning can strike things in the air or on the ground. And contrary to what many people think, it can strike the same place twice. Bolts often hit tall buildings hundreds of times. One day, they hit the Empire State Building in New York City, United States, 15 times in 15 minutes.

In the Arctic and over oceans, lightning is less common. But every second, 100 to 200 bolts hit Earth. A large bolt can contain

Mighty God of Thunder

About 2000 years ago in Scandinavian countries, such as Sweden and Norway, Vikings told stories about Thor, god of thunder. Big and strong, he was one of the mightiest gods who lived on a mountaintop in the center of the universe. When he rode his chariot, hills shook, lightning flashed and thunder rolled. When he used his magic hammer, the skies boomed.

Thor had a quick temper. If he was angry, everyone feared him. But most of the time, he was good-natured and kind. Thor used his strength and his magic hammer mostly to protect other gods from evil giants.

100 million volts of power. Lightning often causes fires in forests and power blackouts in cities. And it kills more people than any other weather force. That's why, during thunderstorms, it's important not to stand out in the open or under a lone tree where you might attract lightning.

Stone Cold

Some thunderstorms produce hailstones—white balls or cones of ice that grow in layers like an onion. They begin as tiny water droplets high up in a cloud. The droplets freeze and start to fall, but air currents in the thundercloud often carry them back up. The droplets gather more droplets, which also freeze. They can fall and rise several times, forming more layers of ice. When they grow too big and heavy to rise in the cloud, they fall to Earth as hail.

Each hailstone contains up to 25 layers and billions of water droplets. Sometimes the air currents that flow upwards carry leaves and pebbles to the clouds. There they mix with the droplets and freeze inside the hailstones. Some stones even contain small animals, like insects, tiny fish or turtles that are swept into the clouds.

Hailstones are usually the size of marbles. Some grow to be as big as golf balls or baseballs. But even larger stones fell in Kansas, United States, in 1970. Some of them weighed three-quarters of a kilogram (just under 2 pounds) and clobbered the ground while

Hail Brought Peace

In France during 1359, England's King Edward III was preparing his army to attack. Suddenly, a powerful thunderstorm broke loose. Giant hailstones bombed the troops, killing thousands of soldiers and their horses. Some of the soldiers wore thick metal helmets, which protected their heads. But the hailstones grew even larger and killed many of these soldiers as well. Those who lived through the storm were terrified—and so was King Edward. Rather than continue the attack, he decided to seek peace with France. Soon after, the two countries signed a treaty.

People have tried to shatter hail before it fell by firing cannons loaded with gunpowder.

traveling 160 kilometres (100 miles) an hour. People in Bangladesh in 1986 reported hailstones that weighed a kilogram (over 2 pounds).

Not surprisingly, hail crushes crops, strips bark off trees, dents cars and smashes windows. Sometimes, it even kills animals and people. Most hailstorms last just a few minutes, but one in Kansas went on for more than an hour, dumping a layer of stones 45 centimetres (18 inches) deep.

Surprisingly, some of the places that hail strikes most often lie close to the equator. The high hills of Kenya are hail-hit about 130 days a year.

* * *

Thank heaven for thunderstorms. They are one of the forces that shift heat from ground to air and spread moisture around the planet. And by drawing in cold drafts, thunderstorms help to clear the air. Even lightning does a lot of good. It helps unite elements in the air, like nitrogen and oxygen, that combine with water droplets to produce half the natural fertilizer on Earth.

Great Balls of Fire

Flying through a dark cloud, the pilot of a small plane was startled by lightning. Unlike any he had seen before, it was the size and shape of a tennis ball. The lightning bounced into the plane through an open window in the cockpit. It scorched the pilot's eyebrows, then rolled to the back of the plane where it suddenly exploded. Although it made a loud bang, it caused no harm.

For centuries, thousands of people have spotted ball lightning like this. During thunderstorms, they have seen it rolling along the wings of planes, crossing grassy yards—even circling inside homes.

Some scientists think thunderstorms cause ball lightning by breaking down air and building up a hot core of electrical charges. Layers develop around the core, forming a fiery ball with a balloonlike surface.

The glowing balls vary in color and brightness. Traveling up to 10 metres (33 feet) a second, they usually last less than a minute. Then they explode or fade away.

INDEX

Aeolus 38
Africa 18, 28, 33
Aftershocks 7
Air currents 51
Alaska 5, 10, 29, 31
Alberta 44, 45
Amazon River 17
Antarctica 11, 29
Arctic 49
Asia 7, 10
Atlantic Ocean 18, 25, 27, 29, 33
Australia 33
Avalanches 1, 8, 9–10, 13

Ball lightning 53
Bangladesh 53
Bering Glacier 10
Black smokers 23–25
British Columbia 9

Canada 9, 23, 44, 45
Caribbean Sea 29

Chile 7, 33
Chinook Arch 45
Chinooks 42, 43–45
Clouds 38, 39, 40, 43, 45, 48, 51
Cone geyser 21
Cyclones 38
Cyprus 6

Earthquake light 7
Earthquakes 1, 2, 3–9, 10, 13, 18, 27, 28, 29, 30, 31, 33, 35
Edrie 31
Electricity 1, 13, 19, 25, 35, 47, 48, 53
Empire State Building 49
England 28, 52
Europe 7, 17
Eye, of hurricane 38, 39, 40
  of tornado 40

Fires 7, 51

Floods 7, 17, 31, 38, 43
Foehns 42, 43–45
Fork lightning 48
Fountain geyser 21
France 52

Geysers 15, 17, 19–23, 25
Glaciers 10, 11–13, 17
Gohei 32–33
Great Geysir 19
Great Meteor Seamount 18
Greeks 38
Gulf of Mexico 41

Hail 41, 47, 51–53
Hawaii 17, 29, 31, 35
Hilo Bay 35
Himalaya Mountains 10
Hiro 32
Hoover Dam 5

Hurricane hunters 40
Hurricanes 28, 37, 38–40, 45
Hydrothermal vents 23–25

Ice 1, 3, 9, 10, 11–13, 17, 43, 51
Iceland 17, 18, 19, 21
India 4, 5
Indian Ocean 33
Indonesia 16, 18, 33
International Decade of Natural Disaster Reduction 2
Io 17

Japan 2, 7, 21, 27, 28, 32, 35
Jupiter 17

Kansas 43, 51, 53
Kariba Dam 5
Keller, Will 43

Kenya 53
King Edward III 52
Kourion 6–7
Krakatoa 33

Landslides 7, 16, 28, 29, 33
La Plata 29
Lava 16, 17, 18, 19, 33
Lightning 1, 18, 43, 47–51, 53
Lituya Bay 31

Magma 15, 18, 25
Mars 17
Meteorites 5, 28, 29
Mount Pinatubo 18
Mount Tambora 16
Muha-pudma 4

New York City 49
New Zealand 21
North America 10, 17, 29
Norway 50

Ocean currents 28
Old Faithful 21
Olympus Mons 17

Pacific Ocean 5, 7, 17, 23, 25, 27, 28, 31, 35, 45
Pacific Rim 31, 33
Peru 4, 10
Philippines 17, 18, 35
Puerto Rico 38

Rain 38, 40, 41, 43, 47
Richter scale 5, 7, 9, 29
Rogers Pass 9
Rogue waves 28
Romans 4
Russia 5, 16, 21

Saffir-Simpson scale 40
Scandinavia 50
Scotland 27
Seamounts 18
Sheet lightning 48
Snow 3, 9–10, 11, 13, 17, 43, 44–45
South America 23
South Dakota 45
Steam 15, 18, 19, 21, 23, 25, 33
Surtsey 18
Sweden 50
Switzerland 11

Texas 29

Thor 50
Thunder 47, 48, 50
Thunderstorms 40, 47–53
Tidal waves 28
Tornado Alley 41
Tornadoes 2, 36, 40–43, 45
Trabant, Dennis 10
Tsunamis 27, 28–35
Tsunami Warning System in the Pacific 35
Twisters 40–43
Typhoons 38

Ulrich, Sonny 31
United Nations 2
United States 5, 13, 18, 21, 22, 29, 41, 45, 49, 51
Unteraar Glacier 11
U.S. Geologic Survey 10
USS *Ramapo* 28

Vents 16, 18, 19, 23–25
Vikings 50
Volcanoes 1, 2, 5, 10, 15–19, 21, 25, 27, 28, 29, 31, 33, 48
Volcano Katla 17

Waimangu Geyser 21
Wakayama 32
Warning systems 2, 17, 18, 35, 40, 43
Waterspouts 41
Waves 1, 2, 7, 27–35
Wilson, Kerry 23
Winds 1, 2, 10, 27, 28, 36–45
Wyoming 22

Yellowstone National Park 22

Zambia 5

55

ABOUT THE AUTHOR

Diane Swanson lives on Vancouver Island, British Columbia. Her articles on nature and wildlife have appeared in children's magazines *Ranger Rick* and *Owl*. She is the author of five "Our Choice" children's books: *A Toothy Tongue and One Long Foot*, *Why Seals Blow Their Noses*, *Squirts and Snails and Skinny Green Tails*, *Safari Beneath the Sea*, and *Coyotes in the Crosswalk*. Diane Swanson is also the author of *Sky Dancers* and *The Emerald Sea*.

ABOUT THE ILLUSTRATOR

Laura Cook is a freelance illustrator who lives in Vancouver, British Columbia. She worked as a registered nurse for eight years before turning to art full time. Her work has been featured in publications such as *Bon Appetit*, *Eating Well*, and *Vermont Magazine*. Her first book is a collection of recipes with whimsical drawings entitled *Feasts of Fancy*.